What were schools like in early America? They were very different from today!

There were no schools at first. Young people learned at home. Their mothers and fathers were their teachers.

The first schools were still in a home. A housewife was the teacher. She often did her own work too. She might knit while students were studying!

When schools were finally built, they were only one room. There was one teacher for all ages. Students learned to read, write, and spell. They also worked with numbers.

Some teachers were strict and impatient. Students knew they should not cause trouble.

These early schools had very few books. Some schools only had hornbooks. These were not like books with pages. They were usually small wooden paddles. A piece of paper was tacked to the wood. A slice of a cow's horn went over the paper. It was thin enough to read through.

If schools had books, they were usually "primers." The class would copy the primers over and over. Students had to recite the lessons. It was crucial that they know the right answers to questions.

Along with very few books, there was very little paper. Some students brought small chalkboard slates to school. Other students wrote on tree bark. They would use chalk and lead pencils.

Pens could be made from wood or feathers. Those made from feathers were called quill pens.

Ink was made from crushed roots and nuts. There was berry ink too. The inks could be beautiful colors. They might be blue, red, brown, or gold.

Sometimes students lined up for a spelling bee. This exercise tested spelling skills. People still have spelling bees today!

Imagine school life in early America. Would you like to have been a student at that time?

Fantasy £2

HEINEMANN
START

STEPHEN C

The Arcade

HEINEMANN

ARCADE
OPEN UNTIL MIDNIGHT

It is late. But the arcade is open. Everybody is in the arcade. They are watching Freddy. Freddy is playing Time Killer.

3

TIME ⧖ KILLER

1 DOLLAR TO PLAY

11:45:00

SCORE 12,000,00 POINTS TO WIN!

LAST GAM

SCORE 10,999,99 LEVEL 11 WINNER TIME LOSER

I am Time. Welcome to Time Killer. Win a prize! It's a big surprise!

There are twelve levels to my game. Can you score twelve million points? Can you kill Time? Can you win?

Please put one dollar in the machine.
Welcome to Level One.
The time is now eleven forty-five.
It's late and I don't wait.

Freddy, you come here every day.
You play this game all day.
You're wasting your time. You're wasting your money. You always lose.

Stop playing. It's time to go home.
It's nearly midnight.

I can get to Level Twelve. That's the end of the game.

The winner gets a prize. I'm going to win. I'm going to enter my name in the machine. I'm going to be the winner. I'm going to kill Time!

Now shut up, Pete! Go away!

You're wasting your time, Freddy.
You can't win. The machine always wins. You always lose.

Look at the time. It's nearly midnight.
The arcade closes at midnight.
It's time to go home.

Stop playing. Let's go.

TIME ⧖ KILLER

1 DOLLAR TO PLAY

11:57:00

SCORE 12,000,000 POINTS TO WIN!

LAST GAME

SCORE 11,999,999
LEVEL 12
WINNER TIME
LOSER ★

Time wins. You lose.
Do you want to play again?

Please put one dollar in the machine.

It's three minutes to midnight.
It's late and I don't wait.

Look at my score. It's nearly twelve million. I can win! I can win! I must play again. I must play one more time. Nothing can stop me.

But I haven't got any more money. Give me a dollar. I'm going to play one more time.

No. I'm not going to give you any money. It's midnight. The arcade is closing now. I'm going home. Are you coming?

Pete leaves the arcade. He wa[lks] the alley. Freddy follows him. It is dark. It is late.

Pete, give me some money. I must play one more time. I must! Give me a dollar.

No. I'm not going to give you any money.

THE ARCADE →

Freddy hits Pete. Pete falls to the ground. Freddy takes some money out of Pete's pocket.

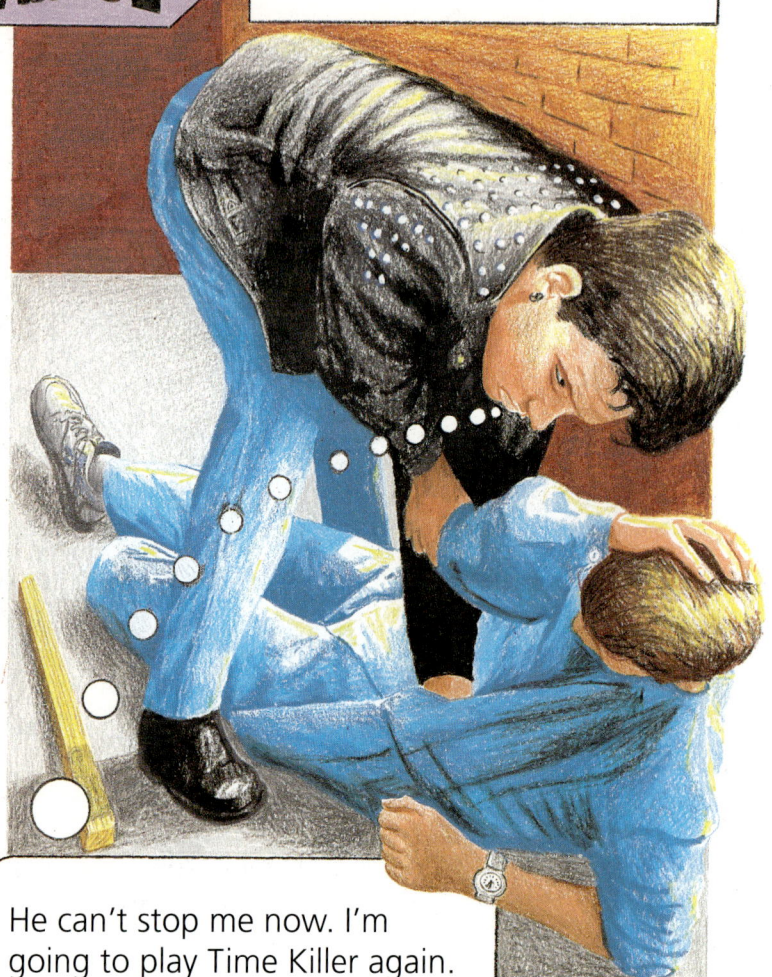

He can't stop me now. I'm going to play Time Killer again. I'm going to finish Level Twelve. I'm going to win this time.

OPEN UNTIL MIDNIGHT

I'm on Level Twelve.
My score is nearly twelve million.
I'm going to win! I'll kill Time.

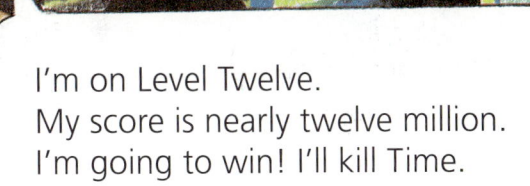

Then I'll enter my name in the machine. I'm going to get the prize!

I win! I'm at the end of Level Twelve. Time is dead. I can enter my name in the machine. I'm the winner! Time is the loser.

TIME KILLER

1 DOLLAR TO PLAY

13:00:00

SCORE 12,000,000 POINTS TO WIN!

LAST GAME
SCORE
00:00
LEVEL 13
WINNER ★
LOSER
TIME CAN NEVER LOSE

Congratulations. You are the new winner. Welcome to Level Thirteen.
Please enter your name.
The time is thirteen o'clock.

I'm the winner! I'm entering my name. F-R-E-D-D-Y. Enter.

… No-o-o … WAIT!

TIME KILLER

1 DOLLAR TO PLAY

01:00:00

SCORE
12,000,00
POINTS
TO WIN!

LAST GAME

SCORE
00:00
LEVEL ★
WINNER
FREDDY
LOSER ★

I am Time. Welcome to Time Killer. Win a prize! It's a big surprise!

There are twelve levels to my game. Can you score twelve million points? Can you kill Time? Can you win?

Please put one dollar in the machine. Welcome to Level One. The time is now one o'clock. It's late and I don't wait.